Newton & Descartes's Coolest, Rockin' Day Ever

MATH MUSICALS

www.MathMusicals.com

Volume 1

by
Anne Lazo
and
Michael Wiskar

BIG IDEAS LEARNING®

Erie, Pennsylvania
BigIdeasLearning.com

Cartoon Illustrations: MoreFrames Animation
Photo of Michael Wiskar provided by Greg Blue *www.gregblue.com*

Stockphoto credits:
MimaCZ/iStock/Getty Images Plus
dimonspace/iStock/Getty Images Plus
ArtyCool/iStock/Getty Images Plus
Mjak/iStock/Getty Images Plus
rvika/iStock/Getty Images Plus
Goldmund/iStock/Getty Images Plus

Printed in the United States.

ISBN 13: 978-1-63598-898-7

123456789–22 21 20 19 18

Newton   *Descartes*

Contents

The authors of *Big Ideas Math* understand the importance of storytelling and music in learning. Storytelling puts math in real-life contexts, and music helps children remember. In collaboration with storyteller Anne Lazo and musician Michael Wiskar, the authors of *Big Ideas Math* are excited to offer *Math Musicals*—engaging musicals that bring mathematics to life!

In this fictional world, students can interact with many different characters, including Newton and Descartes! Students will often wonder—What adventures will Newton and Descartes have today? What will they learn? What will they do next? Using *Math Musicals* will not only engage students in the mathematics, but will help improve their literacy skills as well.

Math and literacy skills are critically important in this world. Bringing math and literacy together through music is the perfect way to instill a lifelong love of learning, encouraging young minds to love math. At Big Ideas Learning, we believe providing the opportunity for students to interact with and learn the math through storytelling and music is just one more way to make learning visible.

About the Authors

Anne Lazo is an international performing arts educator, theatrical director, and author and composer of children's musical educational books.

Anne graduated from the University of Southern California and has over 20 years experience teaching and managing inspirational arts programs and productions for students, as well as professional actors, musicians, and dancers.

Anne worked in Japan for 15 years as an accomplished musical theater director, acting coach, choreographer, script writer, director of educational programs, and motivational speaker for arts advocacy. She has worked for Disney's World of English (World Family English), NHK Television and Radio, and owned her own production company, where she produced and directed musicals for schools and companies. In the United States, Anne was the Director of Education for the New West Symphony, where she developed and implemented many community outreach programs for all ages throughout the Los Angeles area.

Anne currently lives with her husband in Spain and is the Creator and Founder of My Child and Me - English©. Anne and her theatrical team of actors and teachers provide ESL educational shows, musical storybooks, and classes for young children.

Michael (Mike) Wiskar is a Leo-nominated composer, songwriter, independent music producer, vocalist, and multi-instrumentalist.

A graduate of the Music Program at MacEwan University as well as Langara's Digital Music Production Program, Mike began his music career in his teens as a performer in rock, country, and jazz bands before working as a songwriter for various artists in Quebec. There he worked closely with Myles Goodwyn of legendary Canadian rock band April Wine, among other top-level songwriters and producers.

Mike has since written and produced songs and music scores for theater, short films, full length and series documentaries, choirs, contemporary artists, and numerous television series (Operation Vacation, Psycho Kitty, Men's Fashion Insider, Kitty 911, and Favorite Places, to name just a few).

Math Musicals Stories and Topics

Grade K
Newton & Descartes's Coolest, Rockin' Day Ever
- *Fish Crackers* (Counting to 10)
- *Cool Cats & Rockin' Dogs* (Partner Numbers to 5)
- *Big Bulldog Bob* (Counting to 20)
- *JAM Session* (Naming Two- and Three-Dimensional Shapes)
- *Best Friends (I Get You and You Get Me)*

JAM Session

Grade 1
Newton & Descartes's Day at the Beach
- *Seashells* (Addition to 10)
- *Cora's Home* (Subtraction to 10)
- *100 Waves* (Counting to 100 by 10s)
- *Racing the Clock* (Telling Time)
- *Best Friends (I Get You and You Get Me)*

100 Waves

Grade 2
Newton & Descartes's Four-Legged Fun
- *Four-Legged Games* (Two-Digit Addition)
- *Newton & Descartes's Turn* (Two-Digit Addition and Subtraction)
- *Pet Airlines* (Counting to 1,000 by 100s)
- *At the Airport* (Money)
- *Best Friends (I Get You and You Get Me)*

At the Airport

Grade 3
Newton & Descartes's Night in Madrid

- *Tiny Tapas* (Multiplying by 3)
- *Flamenco Mice* (2s, 3s, 4s, and 5s Multiplication Facts)
- *Where is the Park?* (Fractions)
- *Fútbol Fun* (Perimeter)
- *Best Friends (I Get You and You Get Me)*

Fútbol Fun

Newton Leads the Way

Grade 4
Newton & Descartes's Skate Park Adventure

- *Jump Start* (Estimating Sums and Products)
- *The Amphibian* (Divisibility Rules for 2, 3, and 5)
- *The Catwalk* (Adding and Subtracting Fractions with Like Denominators)
- *Newton Leads the Way* (Types of Measurement)
- *Best Friends (I Get You and You Get Me)*

Grade 5
Newton & Descartes's Pet Center Adventure

- *Happy Paws Club* (The Language of Decimals)
- *The Cat's Eye Club* (Multiplying Decimals and Whole Numbers)
- *Bark Like a Dog* (Adding Fractions with Unlike Denominators)
- *Friends* (Volume of a Rectangular Prism)
- *Best Friends (I Get You and You Get Me)*

Friends

Accompanying Products

Visit *www.MathMusicals.com* for the following products.

Songs

Math Musicals songs are available for use throughout the year. Play the songs from *Math Musicals* to not only help students learn concepts, but to help them review all year long as well!

Animated Videos

The *Math Musicals* songs and storybooks come to life in these engaging animations filled with vivid and fun images that your students are sure to enjoy. Join Newton and Descartes on adventures around the world! From rocking out at a jam party to flying high in an airplane, Newton and Descartes engage students in mathematics in fun and innovative ways!

Sheet Music

Each *Math Musical* song comes with its own sheet music. Students and teachers can engage in learning by playing along with songs from *Math Musicals* on a piano or a guitar! Teachers have the ability to truly mix the teaching of math and music with this easy-to-read sheet music.

Differentiated Rich Math Tasks

Differentiated Rich Math Tasks are a fun and intriguing way for students to interact with mathematics. These leveled worksheets challenge students to stretch their knowledge and dig deeper! Each task has three different levels so teachers can easily differentiate to meet the needs of every student.

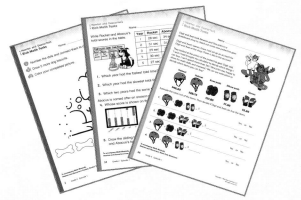

Puppets and Plush Toys

You can use the Newton and Descartes puppets and plush toys to make learning fun and to help students talk about math!

Newton & Descartes's Coolest, Rockin' Day Ever

by
Anne Lazo and Michael Wiskar

Fish Crackers (Counting to 10)
Cool Cats & Rockin' Dogs (Partner Numbers to 5)
Big Bulldog Bob (Counting to 20)
JAM Session (Naming Two- and Three-Dimensional Shapes)

Newton and Descartes's owner, Miss Polly, got ready to walk out the door to go to work. She patted her pets on their heads and said, "Goodbye Newton, goodbye Descartes. Have a good day and please behave! I'll be back after work."

With Polly gone, Newton and Descartes were home alone for the rest of the day. "I guess we won't be playing outside today," said Newton sadly as he slumped down and waited patiently for Polly to return. "Another boring day stuck inside the house."

"I'm not going to wait by the door all day," said Descartes, as he ran off to look for a treat in the kitchen. "I'm hungry!"

Suddenly Descartes remembered that there was a box of fish crackers, just above the counter, in the cabinet. He swiftly jumped up, found his way into the cabinet, and sighed with relief.

"Oh, here they are! Yes…. excellent! It looks like there are still some left, and they're all mine!"

Descartes opened the box of crackers and gently dumped them onto the floor. "How many fish crackers are in the box?"

"1-2-3-4-5-6-7-8-9-10 fish crackers," counted Descartes. "10 fish crackers and they're all mine!"

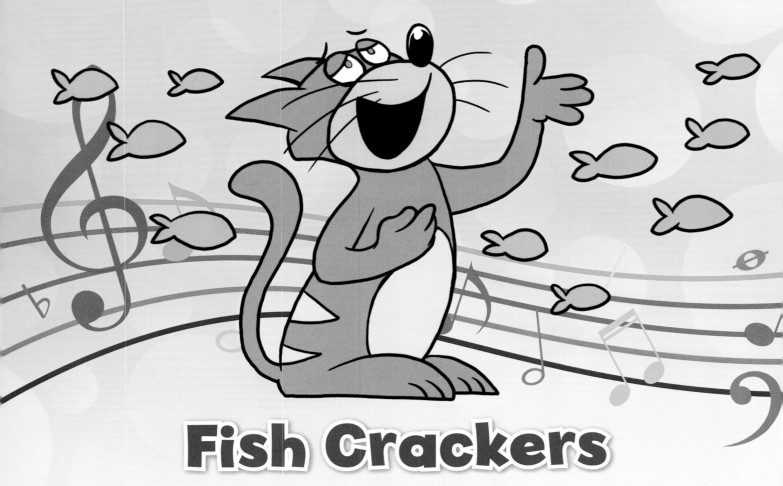

Fish Crackers

by Michael Wiskar

I've got fish crackers 1-2-3, fish crackers and they're all for me.
I've got fish crackers 4-5-6, fish crackers, in the mix.
I've got fish crackers 7-8-9, fish crackers and they're all mine.
I've got 1-2-3-4-5-6-7-8-9 ... 10. Let's do it again.

I've got fish crackers 1-2-3, fish crackers and they're all for me.
I've got fish crackers 4-5-6, fish crackers, in the mix.
I've got fish crackers 7-8-9, fish crackers and they're all mine.
I've got 1-2-3-4-5-6-7-8-9 ... 10. Let's do it again.

1-2-3-4-5-6-7-8-9 ... 10. Fish crackers!
1-2-3-4-5-6-7-8-9 ... 10. Fish crackers!
1-2-3-4-5-6-7-8-9 ... 10. Let's do it again!

1-2-3-4-5-6-7-8-9 ... 10. Fish crackers!
1-2-3-4-5-6-7-8-9 ... 10. Fish crackers!
1-2-3-4-5-6-7-8-9 ... 10. Fish crackers!

1-2-3-4-5-6-7-8-9 ... 10 and they're all mine!

Newton perked up his ears and sniffed the air. He could hear the sound of crunching, so he raced over to the kitchen to see what Descartes was munching on.

"Hey, what are you eating?" asked Newton. "Did you save some for me?"

Descartes was licking the last crumbs off of his paw as he replied, "Well, I was going to call you, but you wanted to wait by the door and I know you don't like fish crackers!"

"I'm hungry too," barked Newton. "Why didn't you share those crackers with me?"

"Don't worry," said Descartes, "I was only eating what *cats* like to eat. I think your *dog* biscuits are in this jar."

Newton was so excited to see the biscuits, but his big doggy paws knocked the jar off the counter and onto the floor.

"Oh no! Newton, be careful!" warned Descartes.

"Dog biscuits! Dog biscuits! Oh yeah, I have my own biscuits!" Newton yelped with excitement as his tail wagged briskly through all of the biscuits that fell out of the jar and onto the kitchen floor.

Descartes just shook his head and said, "Well there *were* 10 nice biscuits in the jar. Now there are 10 biscuits all over the floor and they're all for you!"

"Wow! Look at all of these biscuits! 1-2-3-4-5-6-7-8-9-10 biscuits and they're all mine!" said Newton happily as he gobbled them all up.

"I'm glad you are enjoying all of the dog biscuits, Newton. Don't forget to clean up your mess!" said Descartes.

Descartes had cleverly, and ever so neatly, put the empty box of fish crackers back into the cupboard as if he had never touched it. Then, before Newton had finished his last, yummy dog biscuit, Descartes handed him a broom and dustpan.

"What's this?" Newton asked while slobbering his last morsel of food.

"It's clean up time," said Descartes. "I'll just get out of your way while *you* clean up *your* mess!"

Newton just stared at the mess on the floor and sighed.

COOL CATS & ROCKIN' DOGS

by Anne Lazo

4 and 1 are partner numbers of 5

3 and 2 are partner numbers of 5

Descartes sat on the window ledge in the living room and rested after eating *all* of the fish crackers. Newton cleaned up his mess in the kitchen and dashed into the living room to see what Descartes was doing next.

"Hey, Descartes, what do you want to do now?" asked Newton, still drooling from eating so many dog biscuits. "Since Miss Polly isn't coming home until later I thought that maybe we could play a game?"

Descartes looked out the window and heard meowing in the distance. All of a sudden he got an idea and told Newton, "I'm going to call my Cool Cat friends over to play some jazzy jazz music. We'll have a cat jam session!"

"What's a cat jam session?" asked Newton.

"It's when a group of musical cats play cool music together. The jazzy jazz group has 1 Cool Cat that sings like a dream and 4 Cool Cats that play instruments!" replied Descartes excitedly.

Newton thought to himself, *1 Cool Cat that sings and 4 Cool Cats that play instruments? Oh no, how many cats are coming into the house? 1 and 4 is 5! We can't have 5 jazzy jazz, I mean Cool Cats, in the house playing music together!* "DESCARTES WAIT! That's not the kind of game I was thinking of!"

Descartes ignored Newton and poked his head out of the window. He meowed to the left and meowed to the right. He lifted his paws and started meowing to a beat. "Meow, meow, meow. Meow, meow, meow. Let's play music, friends. Meow, meow, meow."

The Cool Cats in the neighborhood heard Descartes meowing a jazzy jazz beat so they meowed too. "Meow, meow, meow. Meow, meow, meow. Let's play music, friends. Meow, meow, meow." One at a time, all **5** Cool Cats entered the house through the living room window from a tree branch outside.

"Say hello to my *cool* friends, Newton," said Descartes.

Newton barked and growled, but the Cool Cats weren't afraid. They set up their instruments and started playing a jazzy jazz beat.

Descartes clicked his claws to the beat and said with a grin, "Newton, listen... Now THIS is a COOL CAT JAM session. Oh yeah!"

Cool Cats

by Michael Wiskar

Cool cats climbing in my window.
4 cats laying down the beat, and
1 on the microphone.
Yeah, she sings really sweet while
　She's tapping her feet.
4 and 1 is 5 cats.
1 and 4 is 5 cats.
Either way it's complete,
　While they're tapping their feet.
Either way it's complete.

Cool cats climbing in my window.
4 cats laying down the beat, and
1 on the microphone.
Yeah, she sings really sweet while
　She's tapping her feet.
4 and 1 is 5 cats.
1 and 4 is 5 cats.
4 and 1 is 5 cats.
1 and 4 is 5 cats.
Either way it's complete,
　While they're tapping their feet.
Either way it's complete.

Other cats in the neighborhood heard the music and came over to the house to listen to the jazzy jazz beat. 4 more cats climbed through the window to join in the fun.

Newton panicked seeing all of the cats entering the house, so he decided to call some of his rockin' dog friends over to have a dog jam session too! He barked out the window, calling for the Rockin' Dogs to come over and play.

The dogs rushed over and tried to enter the house through the front door. They balanced on top of each other until they could reach the door knob. They turned the knob until the door opened. As they fell to the floor in the house, Newton barked again with joy as he introduced his friends to Descartes.

"Descartes, meet the **5** Rockin' Dogs!" said Newton. "The group has **2** howling dogs that can really sing and **3** paw playing hounds that love to rock and roll! **2 and 3 make up this rockin', barking, howling, 5-dog band!**"

"Wow, great, Newton! Look, there are the same number of dogs and cats in each group. There are **5** Cool Cats and **5** Rockin' Dogs! That's really neat and so complete!" said Descartes very cheerfully.

"Let's all have a jam session together! Come on, Cool Cats, let's show these Rockin' Dogs how to play a great jazzy jazz beat!"

Newton wagged his tail and said, "Come on, Rockin' Dogs, grab your instruments and let's show these Cool Cats how to rock and roll!"

Rockin' Dogs

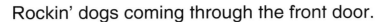

Rockin' dogs coming through the front door.
3 dogs laying down the beat and
2 on the microphone.
Yeah, they're howling so sweet as
    They're rockin' the beat.
3 and 2 is 5 dogs.
2 and 3 is 5 dogs.
Yeah, they're howling so sweet as
    They're rockin' the beat.
Yeah, they're howling so sweet.

and Cool Cats

by Michael Wiskar

Dogs are a rockin' and
  The cats are cooler than cool.
Dogs are a rockin' and
  The cats are cooler than cool.
3 and 2 is 5 dogs.
2 and 3 is 5 dogs.
4 and 1 is 5 cats.
1 and 4 is 5 cats.
Either way it's complete as
  They're rockin' the beat.
Either way it's complete while
  They're tappin' their feet.
Either way it's complete.

4 more dogs entered the house to listen to the rockin', jazzy jazz beat. Newton and Descartes were having a great time with their friends!

"This is the coolest day EVER!" said Newton.

"It sure is! It's the **COOLEST, ROCKIN' day ever**!" said Descartes with a big smile on his face.

BiG BullDog BOB

by Anne Lazo

The Rockin' Dogs and Cool Cats were having so much fun playing their music, wagging their tails, and moving their claws from side to side that they never saw Big Bulldog Bob enter the house.

"What's going on here?" barked Big Bulldog Bob as he slammed the front door with his powerful paws.

Suddenly the music stopped and all of the cats and dogs in the room looked surprised and confused.

"There are some neighbors that are complaining about the noise around here and some pet owners who are looking for their dogs and cats!" growled Big Bulldog Bob as he waved his notepad in the air.

"Who is that?" yelped Newton.

"It's the HOUND from the POUND, Big Bulldog Bob!" declared Descartes in a high-pitched, squeaky voice.

The Rockin' Dogs panicked. They were having a lot of fun and didn't want to be sent home or to the pound! All of the pets quickly ran around the house looking for places to hide. The Rockin' Guitarist crawled behind the couch and Drummer Dog knocked over the garbage can and hid inside.

All of the Cool Cats dashed around the house, jumping up, out of reach of Big Bulldog Bob. It was a real *Jazzy Jazz, Rockin'* mess!

"Where did everyone go?" Newton gasped. "Can you see where everyone went, Descartes?"

"It's no use, I know where all of you are hiding," grumbled Big Bulldog Bob.
"Now come out and get into line so that I can count how many of you are here.
According to my list, there are 18 pets missing!"

He held the whistle that was tied around his neck and blew it as hard as he could.

He kept blowing the whistle very loudly until the Rockin' Guitarist crawled out from behind the couch, Drummer Dog came out of the garbage can, and all of the other dogs rushed into line.

The Cool Cats didn't like the loud sound of the whistle, so they stayed in their hiding places and put their paws over their ears.

"I'm going to count to 10, and when I reach 10, all of the cats hiding in this house had better come into line too!" growled Big Bulldog Bob.

"1-2-3-4-5-6-7-8-9-10."

He pulled out a *fish cracker* box and shook it gently in the air. The Cool Cats didn't come when he blew the whistle, but when they saw the fish crackers, they dashed into line as fast as they could hoping for a treat!

"Now, please get into line with the dogs and stand still while I count how many of you are here.

"1-2-3-4-5-6-7-8-9-10-11-12-13-14-15-16-17-18-19-20.

"20? Hmm."

Big Bulldog Bob

by Michael Wiskar

I'm Big Bulldog Bob, I'm just doing my job.
I'm Big Bulldog Bob, I'm just doing my job.
I'm Big Bulldog Bob, I'm just doing my job.
I round them up, count them up and send them on home.

    1-2-3-4-5-6-7
    8-9-10...11.
    12-13-14-15-16-17-18-19-20.

Hmm now isn't that funny!
I'm Big Bulldog Bob, I'm just doing my job.
I'm Big Bulldog Bob, I'm just doing my job.
I'm Big Bulldog Bob, I'm just doing my job.
I round them up, count them up and send them on home.

    1-2-3-4-5-6-7
    8-9-10...11.
    12-13-14-15-16-17-18-19-20.

Hmm now isn't that funny!
I'm Bulldog Bob, I'm just doing my job.
My name's Bulldog Bob, I'm just doing my job.
My name's Bulldog Bob and what am I doing?
I round them up, count them up and send them on home.
I said, I round them up, count them up and send them on home.
I round them up, count them up and send them on home.

    1-2-3-4-5-6-7
    8-9-10...11.
    12-13-14-15-16-17-18-19-20.

Hmm now isn't that funny!

34

"18 and 2 are 20. So… there are actually 20 pets missing, not 18!" declared Big Bulldog Bob.

Before Descartes and Newton could tell him that they lived in the house, he shook his head in disbelief and wrote down the number 20 on his notepad.

"Alright everyone, stop your meowing, purring, barking, yowling, howling, yipping, and yapping! I'd like all of you to leave your instruments here and follow me to the pound. My list is now complete. I found the 18 missing cats and dogs and now I have 2 more. This is my lucky day!" said Big Bulldog Bob with a proud look on his face.

He blew his whistle one more time and called out, "I round them up, count them up, and send them on home!"

"Descartes, what are we going to do?" yelped Newton!

JAM SESSION

by Anne Lazo

Newton and Descartes did not like having Big Bulldog Bob in their house, telling them what to do, and making all of them go to the pound! Descartes was getting mad. He knew he had to think of a plan quickly to stop Big Bulldog Bob. Descartes is very good at math and he loves music, so he thought of a good reason why it was important that all of their friends were playing music together at their house.

"Excuse me, Big Bulldog Bob. My name is Descartes and this is Newton. This is our house and we are having our friends over for a *JAM* session. We can't leave now," Descartes explained as his voice started to rise in anger.

"Tell me 'kitty cat,' what exactly is a *JAM* session?" growled Big Bulldog Bob.

"Kitty cat? What do you mean calling him a kitty cat! He is not a kitten. He is 5 years old!" Newton barked.

Descartes took a deep breath and began to explain again in a calm voice why they were all playing music together. "JAM stands for Just Amazing Math!" All of the pets nodded their heads in agreement with big smiles on their faces. They all went along with Descartes's *story* so that they would not get in trouble. "We are playing our Jazzy Jazz, Rockin' music and learning about math at the same time!"

"Yes, we are learning a lot about math through our JAM session!" agreed Newton, feeling very proud of Descartes's explanation. "*Just Amazing Math! Just Amazing Math! It's Just an Amazing Math day!*" chanted all of the pets with glee.

"Well, I've never seen cats and dogs play together like this before and what does music have to do with math?" asked Big Bulldog Bob. "You can meow, purr, bark, yowl, howl, yip, and yap all you want. I am still taking all of you to the pound where your owners will pick you up!"

"Please wait, Big Bulldog Bob. Give us a chance. Let me explain," said Descartes. Just then Drummer Dog went to his drum and started to play while Descartes clicked his claws, purred, and rapped to the beat.

Meow, Purr, Bark, Yowl, Howl, Yip, Yap.
Meow, Purr, Bark, Yowl, Howl, Yip, Yap.

We are the Cool Cats and Rockin' Dogs and music is our thing. We like to learn math while we sing! Watch us Big Bulldog Bob as we sing about shapes. You can find them, all over the place!

Join our JAM session and please stay. We are having a JAZZY JAZZ, 'ROCKIN' JAM session day!

Shapes

by Michael Wiskar

■ Shapes
They're everywhere you look
The trick is learning just how
to find them

■ Shapes
A **rectangle** or a **square**
Over here and over there
They're everywhere

Just look at the big bass drum
It's a **circle** if ever I've seen one
And look at the neat guitar
It's a **square** with a **triangle**
carved inside it

Shapes
They're everywhere you look
The trick is learning just how
to find them

Shapes
Like a **cone** or a **sphere**
There are so many here
Can you find them?

Now look around the living room
There's a table the shape of a **cube**
And a **cone**-shaped vase on top.
It doesn't stop
There are shapes all around us!

● Shapes
They're everywhere you look
The trick is learning just how
to find them

▲ Shapes
A **triangle** or a **square**
Or a **cone** or a **sphere**
They're everywhere!

Repeat ■ and ■.
Repeat ● and ▲.

The music was so much fun that soon even Big Bulldog Bob was rockin' to the beat! "Look at the cushion on the chair, can you see that it's a **square**? And the **circles** on this light, golly gee, they're very bright!"

"This is great, Descartes! I like learning math through music!" said Big Bulldog Bob as he jumped up on a **cube**-shaped seat and started moving his feet.

All of the pets were cheering him on. They had never seen a bulldog dance before and Big Bulldog Bob could really dance!

"Yay, Bulldog Bob! Let's sing and dance, dance, dance some more!" chanted all of the pets.

Newton and Descartes were so relieved that Big Bulldog Bob was enjoying the JAM session!

"Sorry I was so hard on you boys." Big Bulldog Bob smiled as he tore up the list from his notepad and flung it on the floor. "This has been my BEST LEARNING DAY EVER! I will make sure that all of the pet owners know that you were having JUST an AMAZING MATH day! We all learned a lot and it was so much fun!"

Suddenly a car came up the driveway and Newton knew that Polly had returned home. "She's home, she's home!" barked Newton as he wagged his tail and ran to the door to see if it was really her. "Wait until Polly sees how much fun we are having today!"

"Newton, we can't let her see all of the dogs and cats in our house!" Descartes meowed nervously.

"Well, it was your idea to invite them in the first place!" Newton reminded Descartes.

"Nevermind that, we have to hurry and get everyone out!" cried Descartes as he grabbed Big Bulldog Bob's whistle and blew it as hard as he could.

The Cool Cats and the Rockin' Dogs stopped playing their music and the JAM session ended as the pets ran back into line one by one. Even Big Bulldog Bob picked up his notepad in a hurry and ran into line, confused by the piercing sound of the whistle.

"We only have about **20** seconds to get every instrument and pet out of the house," meowed Descartes in a panicked voice.

Big Bulldog Bob nodded his head in agreement. "You heard Descartes, let's get moving. We only have **20** seconds! By the time I count from **1** to **10**, I want all of the Cool Cats to jump out the living room window!"

1... 2... 3... 4... 5... 6... 7... 8... 9... 10...

50

All of the Cool Cats jumped out of the living room window and back onto the tree branch outside. "By the time I count from **11** to **20**, I want all of the Rockin' Dogs to rush out of the front door!" ordered Big Bulldog Bob.

11... 12... 13... 14... 15... 16... 17... 18... 19... 20!

The Rockin' Dogs made it out the front door just before Polly walked up to the house from her car.

Newton and Descartes excitedly welcomed Polly home. Big Bulldog Bob gave them a wink and a nod of his head as he snuck out the front door behind them. "Hello Newton, hello Descartes! How was your day?" asked Polly as she petted them while walking into the house and into the kitchen.

Little did she know....that today was their COOLEST ROCKIN' DAY EVER! They were both so happy that they had a great JAM session and that all of their friends came over to play. Newton was happy that Descartes was able to get Big Bulldog Bob excited about math and music. They both enjoyed watching him dance for the first time! He was so funny!

More importantly, they made a new friend to invite to the next Jazzy Jazz, Rockin' JAM session! "Well Newton," whispered Descartes, "we don't have to worry about anything now. All of the pets and instruments are gone, Big Bulldog Bob is happy, and no one will ever know what we were up to today!"

Newton was panting and smiling and they were both feeling great when suddenly they heard Polly shout from the kitchen….

"Okay….who ate all of the fish crackers?!!!……..DESCARTES!"

Best Friends
(I Get You and You Get Me)

by Michael Wiskar

A wink, a nod, a sideways glance,
    a little elbow to the ribs.
We speak a secret language and
    no one else knows what it is.
You know me better than
    I even know myself.
It's a mutual connection
    between us and no one else.

You've got a friend,
Whenever you need a friend.
You've got a friend,
Whenever you need a friend.
I can't pretend
    there's anywhere else I'd rather be,
'Cause I get you and you get me.

And even when there's nothing to do
    we're never feeling bored.
We've had a thousand adventures
    and we'll have a thousand more.
You make me laugh so hard
    I feel like I could cry.
And we pick each other up
    when we're together,
We don't even have to try.

You've got a friend,
Whenever you need a friend.
You've got a friend,
Whenever you need a friend.
I can't pretend
    there's anywhere else I'd rather be,
'Cause you've got a friend in me.

'Cause I get you and you get me.
'Cause I get you and you get me.
We're a special kind of family.

And even if there's trouble brewing
    we always know just what we're doing,
'Cause I've always got your back
    and you've got mine.
And everything's fine!

You've got a friend,
Whenever you need a friend.
You've got a friend,
Whenever you need a friend.
I can't pretend
    there's anywhere else I'd rather be,
'Cause I get you and you get me.

Whenever you need a friend.
    ('Cause I get you and you get me.)
Whenever you need a friend.
I can't pretend
    there's anywhere else I'd rather be,
'Cause you've got a friend in me.

'Cause I get you and you get me.
    (Yeah, I get you!)
'Cause I get you and you get me.
    (I think you get me too.)
We're a special kind of family.

Let's sing it!

You've got a friend,
    ('Cause I get you and you get me.)
Whenever you need a friend.
You've got a friend,
    ('Cause I get you and you get me.)
Whenever you need a friend.
You've got a friend,
    ('Cause I get you and you get me.)
Whenever you need a friend.
We're a special kind of family.

You've got a friend,
    ('Cause I get you and you get me.)
Whenever you need a friend.
You've got a friend,
    ('Cause I get you and you get me.)
Whenever you need a friend.
You've got a friend,
    ('Cause I get you and you get me.)
Whenever you need a friend.
We're a special kind of family.

You've got a friend,
    ('Cause I get you and you get me.)
Whenever you need a friend.
You've got a friend,
    ('Cause I get you and you get me.)
Whenever you need a friend.
You've got a friend,
    ('Cause I get you and you get me.)
Whenever you need a friend.
We're a special kind of family.

Words and Music by
Michael Wiskar

Fish Crackers

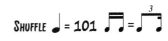

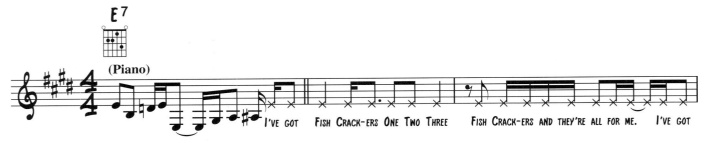

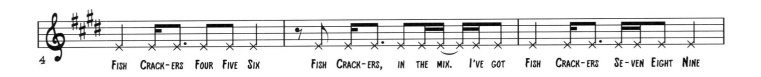

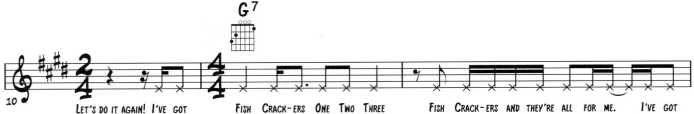

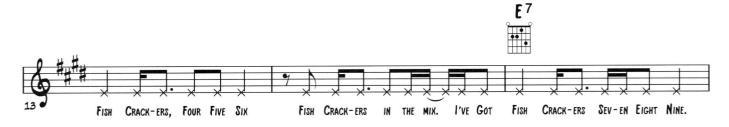

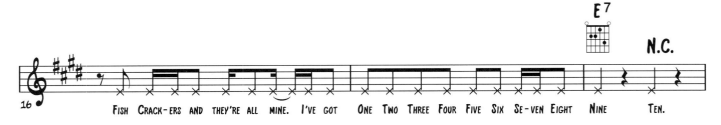

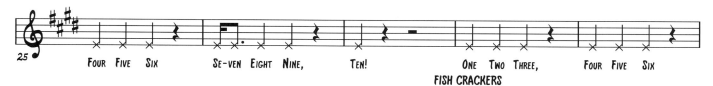

Cool Cats

Words and Music by
Michael Wiskar

Rockin' Dogs and Cool Cats

Words and Music by
Michael Wiskar

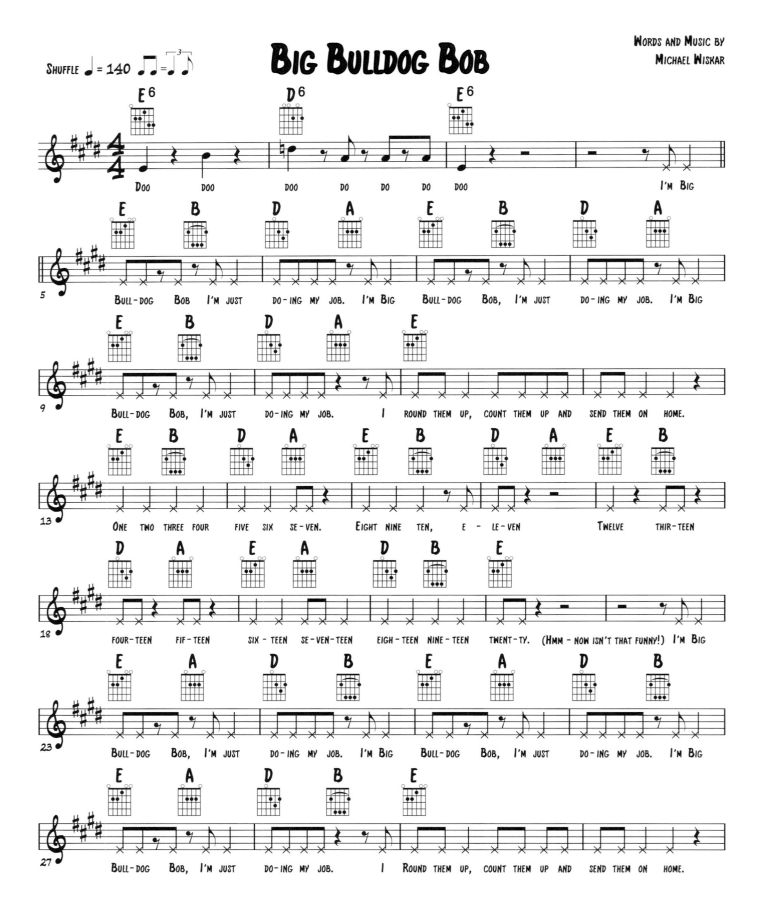

Shapes